ENERGY SECTOR STANDARD
OF THE PEOPLE'S REPUBLIC OF CHINA

中华人民共和国能源行业标准

Code for Design of Densified Biofuel Projects
for Heating Boilers

生物质锅炉供热成型燃料工程设计规范

NB/T 34062-2018

Chief Development Department: China Renewable Energy Engineering Institute
Approval Department: National Energy Administration of the People's Republic of China
Implementation Date: July 1, 2018

China Water & Power Press
中国水利水电出版社
Beijing 2024

All rights reserved. No part of this publication may be reproduced, stored in a retrieval system, or transmitted in any form or by any means—electronic, mechanical, photocopying, recording or otherwise, without prior written permission of the publisher.

图书在版编目（CIP）数据

生物质锅炉供热成型燃料工程设计规范：NB/T 34062-2018 = Code for Design of Densified Biofuel Projects for Heating Boilers (NB/T 34062-2018)：英文 / 国家能源局发布. -- 北京：中国水利水电出版社，2024. 1. -- ISBN 978-7-5226-2719-9

Ⅰ. TK22-65

中国国家版本馆CIP数据核字第2024YU5312号

ENERGY SECTOR STANDARD
OF THE PEOPLE'S REPUBLIC OF CHINA
中华人民共和国能源行业标准

Code for Design of Densified Biofuel Projects
for Heating Boilers
生物质锅炉供热成型燃料工程设计规范
NB/T 34062-2018
（英文版）

Issued by National Energy Administration of the People's Republic of China
国家能源局　发布
Translation organized by China Renewable Energy Engineering Institute
水电水利规划设计总院　组织翻译
Published by China Water & Power Press
中国水利水电出版社　出版发行
　　　　Tel: (+ 86 10) 68545888　68545874
　　　　sales@mwr.gov.cn
　　　　Account name: China Water & Power Press
　　　　Address: No.1, Yuyuantan Nanlu, Haidian District, Beijing 100038, China
　　　　http://www.waterpub.com.cn
中国水利水电出版社微机排版中心　排版
北京中献拓方科技发展有限公司　印刷
184mm×260mm　16开本　2印张　63千字
2024 年 1 月第 1 版　2024 年 1 月第 1 次印刷
Price（定价）：￥280.00

Introduction

This English version is one of China's energy sector standard series in English. Its translation was organized by China Renewable Energy Engineering Institute authorized by National Energy Administration of the People's Republic of China in compliance with relevant procedures and stipulations. This English version was issued by National Energy Administration of the People's Republic of China in Announcement [2023] No. 8 dated December 28, 2023.

This version was translated from the Chinese Standard NB/T 34062-2018, *Code for Design of Densified Biofuel Projects for Heating Boilers*, published by China Water & Power Press. The copyright is reserved by National Energy Administration of the People's Republic of China. In the event of any discrepancy in the implementation, the Chinese version shall prevail.

Many thanks go to the staff from the relevant standard development organizations and those who have provided generous assistance in the translation and review process.

For further improvement of the English version, any comments and suggestions are welcome and should be addressed to:

China Renewable Energy Engineering Institute
No. 2 Beixiaojie, Liupukang, Xicheng District, Beijing 100120, China
Website: www.creei.cn

Translating organization:

Academy of Agricultural Planning and Engineering, MARA

China Renewable Energy Engineering Institute

Translating staff:

MENG Haibo	LI Lijie	AN Zaizhan	FENG Jing
XING Haohan	YE Bingnan	LIU Huan	SHEN Xiuli
WEN Fengrui			

Review panel members:

QIAO Peng	POWERCHINA Northwest Engineering Corporation Limited
LI Zhongjie	POWERCHINA Northwest Engineering Corporation Limited
YAN Wenjun	Army Academy of Armored Forces, PLA

QIE Chunsheng	Senior English Translator
GUO Jie	POWERCHINA Beijing Engineering Corporation Limited
CHE Zhenying	IBF Technologies Co., Ltd.
CONG Hongbin	Academy of Agricultural Planning and Engineering, MARA

National Energy Administration of the People's Republic of China

翻译出版说明

本译本为国家能源局委托水电水利规划设计总院按照有关程序和规定，统一组织翻译的能源行业标准英文版系列译本之一。2023年12月28日，国家能源局以2023年第8号公告予以公布。

本译本是根据中国水利水电出版社出版的《生物质锅炉供热成型燃料工程设计规范》NB/T 34062—2018翻译的，著作权归国家能源局所有。在使用过程中，如出现异议，以中文版为准。

本译本在翻译和审核过程中，本标准编制单位及编制组有关成员给予了积极协助。

为不断提高本译本的质量，欢迎使用者提出意见和建议，并反馈给水电水利规划设计总院。

地址：北京市西城区六铺炕北小街2号
邮编：100120
网址：www.creei.cn

本译本翻译单位：农业农村部规划设计研究院
　　　　　　　　水电水利规划设计总院

本译本翻译人员：孟海波　李丽洁　安再展　冯　晶
　　　　　　　　邢浩翰　叶炳南　刘　欢　沈秀丽
　　　　　　　　温冯睿

本译本审核人员：
　　乔　鹏　中国电建集团西北勘测设计研究院有限公司
　　李仲杰　中国电建集团西北勘测设计研究院有限公司
　　闫文军　中国人民解放军陆军装甲兵学院
　　郄春生　英语高级翻译
　　郭　洁　中国电建集团北京勘测设计研究院有限公司
　　车振英　一百分信息技术有限公司
　　丛宏斌　农业农村部规划设计研究院

国家能源局

Announcement of National Energy Administration of the People's Republic of China
[2018] No. 4

According to the requirements of Document GNJKJ [2009] No. 52, "Notice on Releasing the Energy Sector Standardization Administration Regulations (*tentative*) and detailed implementation rules issued by National Energy Administration of the People's Republic of China", 168 sector standards such as *Guide for Evaluation of Vibration Condition for Wind Turbines*, including 56 energy standards (NB) and 112 electric power standards (DL), are issued by National Energy Administration of the People's Republic of China after due review and approval.

Attachment: Directory of Sector Standards

National Energy Administration of the People's Republic of China

April 3, 2018

Attachment:

Directory of Sector Standards

Serial number	Standard No.	Title	Replaced standard No.	Adopted international standard No.	Approval date	Implementation date
...						
24	NB/T 34062-2018	Code for Design of Densified Biofuel Projects for Heating Boilers			2018-04-03	2018-07-01
...						

Foreword

According to the requirements of Document GNKJ [2015] No. 12 issued by National Energy Administration of the People's Republic of China, "Notice on Releasing the Development and Revision Plan of the Second Batch of Energy Sector Standards in 2014", and after extensive investigation, summarization of practical experience, and wide solicitation of opinions, the drafting group has prepared this code.

The main technical contents of this code include: basic requirements, process design, general layout, civil design, electrical engineering, water supply and drainage and HVAC, fire protection, environmental protection and energy saving, labor safety and industrial hygiene.

National Energy Administration of the People's Republic of China is in charge of the administration of this code. China Renewable Energy Engineering Institute has proposed this code, and is responsible for its routine management and the explanation of specific technical contents. Comments and suggestions in the implementation of this code should be addressed to:

China Renewable Energy Engineering Institute
No. 2 Beixiaojie, Liupukang, Xicheng District, Beijing 100120, China

Chief development organizations:

Academy of Agricultural Planning and Engineering, MARA

POWERCHINA Northwest Engineering Corporation Limited

Chief drafting staff:

ZHAO Lixin	MENG Haibo	HUO Lili	YAO Zonglu
CONG Hongbin	ZHAO Kai	YUAN Yanwen	FENG Jing
LUO Juan	REN Yawei	LI Lijie	DAI Chen
WANG Guan	FENG Taokang	BEN Xian	ZHANG Peng

Review panel members:

YI Yuechun	XIE Hongwen	ZHOU Fengqi	JIA Zhenhang
SHI Dianlin	LIN Cong	YU Guosheng	YANG Xiaoliang
SHI Shutian	TANG Yongjun	TIAN Xiaoxia	QIU Jinsheng
LIU Qijun	LI Shuangjiang	LIU Yuzhuo	HU Xiaofeng
GAO Lijuan	LI Shisheng		

Contents

1	**General Provisions**	1
2	**Basic Requirements**	2
3	**Process Design**	3
3.1	General Requirements	3
3.2	Raw Material Quality Inspection and Storage	3
3.3	Raw Material Pretreatment	3
3.4	Compression Molding	4
3.5	Dedusting	4
3.6	Cooling Packaging and Product Inspection	4
4	**General Layout**	6
5	**Civil Design**	8
5.1	General Requirements	8
5.2	Production Buildings	8
5.3	Accessory Buildings	8
6	**Electrical Engineering**	10
6.1	Load and Power Supply	10
6.2	Power Distribution and Lighting	10
6.3	Lightning Protection and Earthing Design	10
7	**Water Supply and Drainage and HVAC**	12
7.1	Water Supply	12
7.2	Drainage	12
7.3	Heating	12
7.4	Ventilation and Air Conditioning	12
8	**Fire Protection**	13
9	**Environmental Protection and Energy Saving**	14
9.1	Environmental Protection	14
9.2	Energy Saving	14
10	**Labor Safety and Industrial Hygiene**	15
Explanation of Wording in this Code		16
List of Quoted Standards		17

1 General Provisions

1.0.1 This code is formulated with a view to standardizing the design content and depth of densified biofuel projects for heating boilers and ensuring the quality of projects.

1.0.2 This code is applicable to the design of the construction, renovation, and extension of the densified biofuel projects for heating boilers with an annual production capacity of 10000 t or more.

1.0.3 In addition to this code, the design of densified biofuel projects for heating boilers shall comply with other current relevant standards of China.

2 Basic Requirements

2.0.1 The project construction shall comply with national policies and meet the requirements of the local urban-rural development plan.

2.0.2 The project site shall have favorable external conditions of transportation, power supply, water supply, communication, terrain, etc., and shall be located in the place downwind or crosswind of the annual maximum frequency wind of the residential area.

2.0.3 The project design scale shall consider the distribution of biomass feedstock, available resources, purchase and storage, transportation, project construction and other conditions, as well as the demands of heating boilers.

2.0.4 The project site shall keep a safety distance from inflammables and explosives plants, warehouses, high-voltage transmission lines, civil buildings, etc., and shall meet the requirements of the current national standard GB 50016, *Code for Fire Protection Design of Buildings*.

2.0.5 The project construction shall be conducive to ecology and environmental protection, and meet the clean production requirements.

3 Process Design

3.1 General Requirements

3.1.1 The process design shall meet the requirements of the production process flow, product quality and output of densified biofuel for heating boilers.

3.1.2 The process flow includes raw material quality inspection and storage, raw material pretreatment, compression molding, dedusting, cooling packaging, warehousing, factory inspection, etc., and shall meet the relevant environmental protection and energy conservation requirements.

3.1.3 The type selection of main equipment shall consider the following factors:

1. Main technical parameters.
2. Production capacity.
3. Product quality.
4. Compatibility.
5. Energy consumption.
6. Lifespan of vulnerable parts.
7. Dust and noise.
8. Operation and maintenance.

3.1.4 Advanced process flows and equipment should be adopted, the equipment layout should be compact and reasonable, and there should be enough space for operation and maintenance.

3.2 Raw Material Quality Inspection and Storage

3.2.1 Material measurement and quality inspection facilities and equipment shall be provided.

3.2.2 The capacity of raw material storage facilities shall satisfy at least 7-d production demand.

3.2.3 Raw materials should be stored in sheds or raw material warehouses, which shall meet the requirements of ventilation, fire prevention, rain and snow protection, dedusting and smooth transportation of raw materials.

3.3 Raw Material Pretreatment

3.3.1 Raw material pretreatment should include impurity clearing, crushing, drying, mixing, etc., which may be appropriately increased or decreased according to the needs of the production process.

3.3.2 Raw material impurity clearing facilities should be set.

3.3.3 The production capacity of the crushing process shall be greater than 1.3 times the molding production capacity.

3.3.4 Depending on the needs of the molding process, drying facilities may be set after raw material crushing, and mixing equipment may be set before compression molding.

3.4 Compression Molding

3.4.1 A storage bin should be set before the molding machine, and the storage bin capacity shall meet the continuous production needs of the molding equipment.

3.4.2 The molding equipment shall meet the requirements of the current sector standard NY/T 1882, *Technical Conditions for Densified Biofuel Molding Equipment*, and shall ensure product quality and production capacity requirements.

3.4.3 The backup molding equipment shall be reasonably configured according to the needs of the production capacity.

3.4.4 The molding equipment shall be arranged reasonably, and multiple sets of equipment should be arranged in parallel.

3.5 Dedusting

3.5.1 The production workshop shall be equipped with a dedusting system, and the dust concentration in the workshop shall meet the requirements of the current national standard GBZ 2, *Occupational Exposure Limits for Hazardous Agents in the Workplace*. The dust emission shall meet the requirements of the current national standards GBZ 1, *Hygienic Standards for the Design of Industrial Enterprises*; and GB 16297, *Integrated Emission Standard of Air Pollutants*.

3.5.2 The process equipment dispersing dust should be hermetic, and the dust-prone area shall be provided with a suction hood to trap the dust.

3.5.3 Measures shall be taken to prevent secondary pollution of the dedusting system.

3.6 Cooling Packaging and Product Inspection

3.6.1 Air cooling or other cooling methods shall be selected according to the process requirements.

3.6.2 Packaging equipment should be set after the measurement process, and bulk storage tanks may be provided.

3.6.3 Product testing laboratory and relevant testing equipment should be set.

4 General Layout

4.0.1 The general layout shall meet the following requirements:

1. The requirements of the TOR for design and the urban-rural development planning department for the selected plant site.

2. The current national standard GB 50187, *Code for Design of General Layout of Industrial Enterprises*.

3. Make the general layout plan in a unified manner according to the project scale, taking into account the follow-up projects.

4. The layout of the main and auxiliary buildings is reasonable, the production line is compact and standardized, and the flows of people and materials are separated.

5. Compact and reasonable layout, and less land occupation and earthwork quantities on the premise of meeting the production requirements.

6. Comprehensive consideration of building layout, transportation, roads, power supply lines, water supply and drainage pipelines, industrial pipelines, fire protection, greening, environmental protection, etc.

4.0.2 According to the production process flow and functional requirements, the plant area should be divided into office area, raw material area, production area and finished product area, and the areas should be closely linked but properly partitioned.

4.0.3 The building layout shall meet the following requirements:

1. The spacing between buildings meets the requirements of fire protection.

2. The buildings have good natural lighting and ventilation conditions.

3. The impact of production noise and dust on the overall environment is considered.

4.0.4 The traffic system in the plant area shall be designed according to the people flow and material flow routes and the total plant traffic. The roads in the plant area shall meet the following requirements:

1. The roads in the plant area shall be reasonably arranged to facilitate the passage of vehicles. The roads for the transportation of raw materials, finished products and main people flow should use two-way lanes.

2. The factory entrances are reasonably arranged according to the urban-

rural road planning and the principle of separating people flow from materials flow, and there are at least 2 entrances.

3 Fire lanes and emergency exits with obvious signs shall be set.

4 The roads in the plant area should have paved surfaces.

5 Civil Design

5.1 General Requirements

5.1.1 The civil design shall be carried out according to the production process, considering the specific conditions of production scale, site, materials and construction, etc.

5.1.2 The classification of fire hazards for production and auxiliary buildings shall comply with the current national standard GB 50016, *Code for Fire Protection Design of Buildings*.

5.2 Production Buildings

5.2.1 Production buildings should include raw material warehouse, production workshop, finished product warehouse, etc.

5.2.2 Production buildings shall be arranged in zones, and may be partially connected on the premise of meeting the relevant requirements for fire protection and evacuation.

5.2.3 The width, span and height of the production building shall be determined according to the production scale and process.

5.2.4 The indoor floor elevation of the production building shall be at least 0.3 m higher than the outdoor ground design elevation.

5.2.5 Production buildings shall be provided with large equipment transport passages.

5.2.6 The floor of the production building shall be paved, flat, wear-resistant, moisture-proof, and strong enough to meet the production needs.

5.2.7 The live load on the floor of the production building may be calculated by the local load and concentrated load caused by weights involved in production, installation and maintenance, materials stacking, transport vehicles, etc.

5.2.8 A platform and awning for material handling shall be provided at the gates of the finished product warehouse.

5.3 Accessory Buildings

5.3.1 A control room shall be set for the production workshop according to the needs of the equipment, and its location shall meet the following requirements:

    1   Adjacent to the production workshop for easy observation, operation

and dispatching.

2   Good ventilation and lighting, avoiding direct sunlight and glare.

3   Less impact by vibration; the noise shall not be higher than 70 dB.

5.3.2   The design of the control room shall meet the following requirements:

1   The fire endurance rating shall not be lower than Class 2.

2   The indoor clear height shall not be less than 3 m, and the raised floor shall be treated for fire prevention.

3   The computer control room temperature shall be in accordance with the current national standard GB/T 17214.1, *Industrial-Process Measurement and Control Equipment—Operating Conditions Part 1: Climatic Conditions*.

4   Cables shall be accessible for repair.

5   The floor should be made of insulation materials, and antistatic measures shall be taken.

5.3.3   The testing room shall meet the needs for raw materials and finished products testing.

5.3.4   The office area shall keep an appropriate distance from the production area to avoid the effects of noise, vibration and dust.

5.3.5   Separate electromechanical equipment maintenance workshop and spare parts storage shall be set in the plant area.

6 Electrical Engineering

6.1 Load and Power Supply

6.1.1 The production power load shall be of third-grade load, and should be calculated by the demand coefficient method.

6.1.2 The low-voltage distribution circuits shall be differentiated according to the production workshop, production line, work section, work process or units.

6.1.3 A standby circuit with manual switching may be provided for drying, compression molding, and dedusting equipment.

6.1.4 The lighting in the production workshop should be powered by a dedicated line from a low-voltage distribution panel. When it is shared with the power distribution, a main lighting switch and distribution cabinet separate from the power distribution should be set after entering the production workshop.

6.2 Power Distribution and Lighting

6.2.1 The IP level of motors and electrical appliances shall meet the regulations of the current national standard GB/T 4942.1, *Degrees of Protection Provided by the Integral Design of Rotating Electrical Machines (IP code) - Classification*.

6.2.2 The electrical outlets in the production workshop should adopt independent circuit and be equipped with explosion-proof circuit breaker.

6.2.3 When cabling, the filling coefficient of the cable in cable trays, ducts and conduits should not be larger than 40 %. The influence of temperature rise on the safe carrying capacity and the convenience of installation and maintenance shall be considered.

6.2.4 The production workshop shall have production lighting, duty lighting and emergency lighting, and the fire pump room and production monitoring area shall have backup lightings.

6.2.5 Efficient and easy-to-maintain dust-proof and explosion-proof lamps shall be used.

6.3 Lightning Protection and Earthing Design

6.3.1 Various earthing protections in the production workshop should share one earthing grid, and the earthing resistance shall not be larger than 1 Ω; the service power supply system should adopt the TN-S or TN-C-S earthing protection system.

6.3.2 An equipotential bonding system shall be used in the production workshop. Measures should be taken to prevent electric shock in the places out of the equipotential action area.

6.3.3 The incoming line switches for drying and dedusting links should be equipped with alarm and leakage protection devices with delay action.

7 Water Supply and Drainage and HVAC

7.1 Water Supply

7.1.1 Two water supply mains shall be installed in the plant area, one of which shall meet the demands of the whole plant production and domestic water, and the other shall meet the fire water demands of the whole plant.

7.1.2 The production water consumption should be 1.10 to 1.15 times the sum of the production water consumptions of all work processes, and the water supply pipe into the plant area shall be equipped with a master flowmeter.

7.2 Drainage

7.2.1 The drainage system of the plant area shall adopt rain and sewage diversion. The drainage shall comply with the current national standard GB 50014, *Code for Design of Outdoor Wastewater Engineering*.

7.2.2 The access roads in the plant area shall have a good drainage system.

7.3 Heating

7.3.1 The heating system may be installed in the production workshop according to local climatic conditions, and the indoor temperature shall meet the production requirements.

7.3.2 The production workshop shall be equipped with radiators which are not easy to gather dust, easy to clean, and resistant to corrosion.

7.4 Ventilation and Air Conditioning

7.4.1 The ventilation system shall comply with the current national standard GBZ 2, *Occupational Exposure Limits for Hazardous Agents in the Workplace*.

7.4.2 The raw material warehouse and finished product warehouse may adopt natural ventilation or forced ventilation.

7.4.3 Cooling measures shall be taken for the production workshop if the temperature exceeds $36\,°C$.

8 Fire Protection

8.0.1 The fire hazard level and fire endurance rating of the main buildings (structures) shall comply with the current national standard GB 50016, *Code for Fire Protection Design of Buildings*.

8.0.2 In the general layout of fire protection, the fire compartments, fire separation distances, and fire lanes of the main buildings (structures) shall comply with the current national standard GB 50016, *Code for Fire Protection Design of Buildings*.

8.0.3 The distance from the farthest workplace in the production workshop to the exit shall meet the national fire safety management regulations, and there should be at least 2 entrances/exits.

8.0.4 When the process conditions permit, proper fire partitions may be set in the fire compartments.

8.0.5 The fire water supply shall have an independent water source, and shall be provided with fire pool, fire booster pump, etc. The capacity of the fire pool shall be determined according to the sum of the indoor and outdoor firefighting water consumptions within a fire duration. The fire duration is taken as 3 h.

8.0.6 The fire water supply pipelines shall form a circular network, which shall be supplied by at least two water pipes. If one pipeline fails, the total capacity of the remaining pipelines shall be sufficient to meet the fire water demand.

8.0.7 The arrangement of indoor and outdoor firefighting equipment shall comply with the current national standard GB 50016, *Code for Fire Protection Design of Buildings*.

8.0.8 The distribution box of the warehouse with fire risk shall be set separately and installed in the warehouse duty room or outside the warehouse. The wires in the warehouse should be laid in steel pipes.

8.0.9 The distribution lines and control circuits for lighting should be laid according to fire compartments.

8.0.10 The main building (structure) shall be provided with fire evacuation signs and an automatic fire alarm system.

8.0.11 Dedusting ducts shall be provided with explosion-proof and pressure-relief measures.

9 Environmental Protection and Energy Saving

9.1 Environmental Protection

9.1.1 The domestic sewage shall be treated up to standard before discharge.

9.1.2 The noise level in the plant area shall comply with the current national standards GB 12348, *Emission Standard for Industrial Enterprises Noise at Boundary*; and GB/T 50087, *Code for Design of Noise Control in Industrial Enterprises*.

9.1.3 Various exhaust emissions in the production process shall meet the relevant national requirements for emission.

9.2 Energy Saving

9.2.1 Products that have been explicitly prohibited or eliminated by the state shall not be used.

9.2.2 The motor energy consumption of the molding equipment shall meet the current national standard GB 18613, *Minimum Allowable Values of Energy Efficiency and Energy Efficiency Grades for Small and Medium Three-Phase Asynchronous Motors*.

9.2.3 The process design plan shall minimize the frequency of start-stops of the production line, reduce the no-load operation time, and ensure the efficient and stable operation of the equipment.

10 Labor Safety and Industrial Hygiene

10.0.1 A safety education room and a lounge shall be set in the plant area, which shall comply with the current national standard GBZ 1, *Hygienic Standards for the Design of Industrial Enterprises*.

10.0.2 The transmission devices and exposed moving parts shall be provided with safety shields and other protection means. For the visible internal moving parts of the equipment, protective baffles must be installed.

10.0.3 The pits must be covered or fenced.

10.0.4 The high-altitude operation area shall be provided with operating and maintenance platforms and railings, or other safety means.

10.0.5 Places, facilities and equipment with safety hazards or safety risks shall be provided with obvious safety signs and markings. The setting of safety signs and markings shall comply with the current national standards GB 2894, *Safety Signs and Guidelines for the Use*; and GB 2893, *Safety Colours*.

10.0.6 The lighting in the plant area shall comply with the current national standard GB 50034, *Standard for Lighting Design of Buildings*.

Explanation of Wording in this Code

1 Words used for different degrees of strictness are explained as follows in order to mark the differences in executing the requirements in this code.

   1) Words denoting a very strict or mandatory requirement:

      "Must" is used for affirmation; "must not" for negation.

   2) Words denoting a strict requirement under normal conditions:

      "Shall" is used for affirmation; "shall not" for negation.

   3) Words denoting a permission of a slight choice or an indication of the most suitable choice when conditions permit:

      "Should" is used for affirmation; "should not" for negation.

   4) "May" is used to express the option available, sometimes with the conditional permit.

2 "Shall meet the requirements of…" or "shall comply with…" is used in this code to indicate that it is necessary to comply with the requirements stipulated in other relative standards and codes.

List of Quoted Standards

GB 2893,	*Safety Colours*
GB 2894,	*Safety Signs and Guidelines for the Use*
GB/T 4942.1,	*Degrees of Protection Provided by the Integral Design of Rotating Electrical Machines (IP code) -Classification*
GB 12348,	*Emission Standard for Industrial Enterprises Noise at Boundary*
GB 16297,	*Integrated Emission Standard of Air Pollutants*
GB/T 17214.1	*Industrial-Process Measurement and Control Equipment— Operating Conditions Part 1: Climatic Conditions*
GB 18613,	*Minimum Allowable Values of Energy Efficiency and Energy Efficiency Grades for Small and Medium Three-Phase Asynchronous Motors*
GB 50014,	*Code for Design of Outdoor Wastewater Engineering*
GB 50016,	*Code for Fire Protection Design of Buildings*
GB 50034,	*Standard for Lighting Design of Buildings*
GB/T 50087,	*Code for Design of Noise Control in Industrial Enterprises*
GB 50187,	*Code for Design of General Layout of Industrial Enterprises*
GBZ 1,	*Hygienic Standards for the Design of Industrial Enterprises*
GBZ 2,	*Occupational Exposure Limits for Hazardous Agents in the Workplace*
NY/T 1882,	*Technical Conditions for Densified Biofuel Molding Equipment*